Pinpoint Math

Student Booklet
Level E

Volume 4
Multiply and Divide

Photo Credits

©iStock International Inc., cover.

Acknowledgements

Content Consultant:

Linda Proudfit, Ph.D.

After earning a B.A. and M.A in Mathematics from the University of Northern Iowa, Linda Proudfit taught junior- and senior-high mathematics in Iowa. Following this, she earned a Ph.D. in Mathematics Education from Indiana University. She currently is Coordinator of Elementary Education and Professor of Mathematics Education at Governors State University in University Park, IL.

Dr. Proudfit has made numerous presentations at professional meetings at the local, state, and national levels. Her main research interests are problem solving and algebraic thinking.

---

www.WrightGroup.com

Copyright © 2009 by Wright Group/McGraw-Hill.

All rights reserved. Except as permitted under the United States Copyright Act, no part of this publication may be reproduced or distributed in any form or by any means, or stored in a database or retrieval system, without the prior written permission from the publisher, unless otherwise indicated.

Printed in USA.

Send all inquiries to:
Wright Group/McGraw-Hill
P.O. Box 812960
Chicago, IL 60681

ISBN 978-1-40-456803-7
MHID 1-40-4568034

2 3 4 5 6 7 8 9 10 RHR 13 12 11 10 09

Contents

Tutorial Chart vii

Volume 4: Multiply and Divide

Topic 9 Use Multiplication to Compute

Topic 9 Introduction 1
Lesson 9-1 Multiply by Multiples of 10 2–4
Lesson 9-2 Estimate Products 5–7
Lesson 9-3 Multiply: Four Digits by One Digit 8–10
Lesson 9-4 Choose a Method and Multiply 11–13
Topic 9 Summary 14
Topic 9 Mixed Review 15

Topic 10 Use Division to Compute

Topic 10 Introduction 16
Lesson 10-1 Divide Multiples of 10 17–19
Lesson 10-2 Estimate Quotients 20–22
Lesson 10-3 Divide by 1-Digit Numbers 23–25
Lesson 10-4 Choose a Method for Division 26–28
Lesson 10-5 Multiplication and Division 29–31
Topic 10 Summary 32
Topic 10 Mixed Review 33

Topic 11 Equations and Inequalities

Topic 11 Introduction 34

Lesson 11-1 Write Expressions for Patterns 35–37

Lesson 11-2 Write Expressions 38–40

Lesson 11-3 Write Equations with Unknowns 41–43

Lesson 11-4 Solve Equations with Unknowns 44–46

Topic 11 Summary 47

Topic 11 Mixed Review 48

Glossary 49

Word Bank 50

Index 52

Objectives

Volume 4: Multiply and Divide
Topic 9 Use Multiplication to Compute

Lesson	Objective	Pages
Topic 9 Introduction	9.1 Multiply multiples of 10. 9.2 Estimate products by rounding factors and using mental math techniques. 9.3 Solve simple problems involving multiplication of multidigit numbers by one-digit numbers. 9.4 Solve problems by using estimation, mental math, or pencil and paper.	1
Lesson 9-1 Multiply by Multiples of 10	9.1 Multiply multiples of 10.	2–4
Lesson 9-2 Estimate Products	9.2 Estimate products by rounding factors and using mental math techniques.	5–7
Lesson 9-3 Multiply: Four Digits by One Digit	9.3 Solve simple problems involving multiplication of multidigit numbers by one-digit numbers.	8–10
Lesson 9-4 Choose a Method and Multiply	9.4 Solve problems by using estimation, mental math, or pencil and paper.	11–13
Topic 9 Summary	Review computation by multiplication.	14
Topic 9 Mixed Review	Maintain concepts and skills.	15

Topic 10 Use Division to Compute

Lesson	Objective	Pages
Topic 10 Introduction	10.1 Solve problems dividing multiples of 10, 100, and 1,000. 10.2 Estimate quotients by rounding numbers and using mental math techniques. 10.3 Solve simple problems involving division of multidigit numbers by 1-digit numbers with and without remainders. 10.5 Solve problems by multiplying or dividing.	16
Lesson 10-1 Divide Multiples of 10	10.1 Solve problems dividing multiples of 10, 100, and 1,000.	17–19
Lesson 10-2 Estimate Quotients	10.2 Estimate quotients by rounding numbers and using mental math techniques.	20–22
Lesson 10-3 Divide by 1-Digit Numbers	10.3 Solve simple problems involving division of multidigit numbers by 1-digit numbers with and without remainders.	23–25
Lesson 10-4 Choose a Method for Division	10.4 Solve division problems by choosing between using estimation, mental math, or pencil and paper to find the quotients.	26–28
Lesson 10-5 Multiplication and Division	10.5 Solve problems by multiplying or dividing.	29–31
Topic 10 Summary	Review computation using division.	32
Topic 10 Mixed Review	Maintain concepts and skills.	33

Topic 11 Equations and Inequalities

Lesson	Objective	Pages
Topic 11 Introduction	**11.1** Record the rule for a pattern as an expression. **11.2** Write expressions for situations that include an unknown quantity. **11.3** Write equations for word problems that include an unknown quantity.	34
Lesson 11-1 Write Expressions for Patterns	**11.1** Record the rule for a pattern as an expression.	35–37
Lesson 11-2 Write Expressions	**11.2** Write expressions for situations that include an unknown quantity.	38–40
Lesson 11-3 Write Equations with Unknowns	**11.3** Write equations for word problems that include an unknown quantity.	41–43
Lesson 11-4 Solve Equations with Unknowns	**11.4** Write and solve simple equations for word problems that include an unknown quantity.	44–46
Topic 11 Summary	Review writing and solving expressions and equations in one variable.	47
Topic 11 Mixed Review	Maintain concepts and skills.	48

Tutorial Guide

Each of the standards listed below has at least one animated tutorial for students to use with the lesson that matches the objective. If you are using the electronic components of *Pinpoint Math*, you will find a complete listing of Tutorial codes and titles when you access them either online or via CD-ROM.

Level E

Standards by topic	Tutorial codes
Volume 4 Multiply and Divide	
Topic 9 Use Multiplication to Compute	
9.1 Multiply multiples of 10.	9a Using the Partial-Products Method to Multiply
9.1 Multiply multiples of 10.	9b Using Multiples of 10, 100, and 1,000 to Multiply and Divide
9.2 Estimate products by rounding factors and using mental math techniques.	9a Using the Partial-Products Method to Multiply
9.2 Estimate products by rounding factors and using mental math techniques.	9c Using the Standard Multiplication Algorithm
9.3 Solve simple problems involving multiplication of multidigit numbers by one-digit numbers.	9d Solving Word Problems, Example B
9.3 Solve simple problems involving multiplication of multidigit numbers by one-digit numbers.	9a Using the Partial-Products Method to Multiply
9.4 Solve problems by using estimation, mental math, or pencil and paper.	9e Choosing a Method to Solve Multiplication and Division Word Problems
Topic 10 Use Division to Compute	
10.1 Solve problems dividing multiples of 10, 100, and 1,000.	10a Using Multiples of 10, 100, and 1,000 to Multiply and Divide
10.2 Estimate quotients by rounding numbers and using mental math techniques.	10b Estimating Quotients by Rounding Numbers
10.3 Solve simple problems involving division of multidigit numbers by 1-digit numbers with and without remainders.	10c Modeling Division
10.3 Solve simple problems involving division of multidigit numbers by 1-digit numbers with and without remainders.	10d Solving Word Problems, Example C
10.3 Solve simple problems involving division of multidigit numbers by 1-digit numbers with and without remainders.	10e Using the Standard Long Division Algorithm, Example A
10.3 Solve simple problems involving division of multidigit numbers by 1-digit numbers with and without remainders.	10f Using the Standard Long Division Algorithm, Example B
10.3 Solve simple problems involving division of multidigit numbers by 1-digit numbers with and without remainders.	10g Using the Standard Long Division Algorithm, Example C
10.4 Solve division problems by choosing between using estimation, mental math, or pencil and paper to find the quotients.	10h Choosing a Method to Solve Multiplication and Division Word Problems
10.5 Solve problems by multiplying or dividing.	10i Solving Word Problems, Example B
10.5 Solve problems by multiplying or dividing.	10d Solving Word Problems, Example C
10.5 Solve problems by multiplying or dividing.	10j Using Multiplication to Check Division
Topic 11 Equations and Inequalities	
11.1 Record the rule for a pattern as an expression.	11a Using Patterns to Solve Word Problems

11.2 Write expressions for situations that include an unknown quantity.	11b Writing Expressions
11.3 Write equations for word problems that include an unknown quantity.	11c Writing Equations
11.4 Write and solve simple equations for word problems that include an unknown quantity.	11d Solving Equations, Example A
11.4 Write and solve simple equations for word problems that include an unknown quantity.	11e Solving Equations, Example B

Topic 9: Use Multiplication to Compute

Topic Introduction

Complete with teacher help if needed.

1. Solve.

 a. $5 \times 3 =$ _____

 b. Complete the pattern.

 $50 \times 3 =$ _____

 $500 \times 3 =$ _____

 $5,000 \times 3 =$ _____

Objective 9.1: Multiply multiples of 10.

2. Estimate 4×78

 a. Estimate: $4 \times 78 \rightarrow 4 \times$ _____.

 b. So, 4×78 is about _____.

Objective 9.2: Estimate products by rounding factors and using mental math techniques.

3. Multiply 14×8.

 a. Break 14 into tens and ones. _____

 b. _____ $\times 8 =$ _____

 c. _____ $\times 8 =$ _____

 d. Add. _____ $+$ _____ $=$ _____

Objective 9.3: Solve simple problems involving multiplication of multidigit numbers by one-digit numbers.

4. Find the product of 15×12.

 a. Write each fact.

 _____

 _____

 b. Add. _____

Objective 9.4: Solve problems by using estimation, mental math, or paper and pencil.

Lesson 9-1 | **Multiply by Multiples of 10**

Model It

Words to Know A **multiple** is the product of a given number and another whole number.

Activity 1

$3 \times 4 = 12$

$3 \times 4\boxed{0} = 120$

$3 \times 4\boxed{0}\boxed{0} = 1{,}2\boxed{0}\boxed{0}$

$3 \times 4{,}\boxed{0}\boxed{0}\boxed{0} = 12{,}\boxed{0}\boxed{0}\boxed{0}$

Practice 1

$6 \times 7 = \underline{}$

$6 \times 7\boxed{0} = \underline{}$

$6 \times 7\boxed{0}\boxed{0} = \underline{}$

$6 \times 7{,}\boxed{0}\boxed{0}\boxed{0} = \underline{}$

Activity 2

$5{,}000 \times 8 = ?$

$5 \times 8 = 40$

$50 \times 8 = 400$

$500 \times 8 = 4{,}000$

$5{,}000 \times 8 = 40{,}000$

The product of $5{,}000 \times 8$ is $40{,}000$.

Practice 2

$5 \times 2{,}000 = ?$

$5 \times 2 = \underline{}$

$5 \times 20 = \underline{}$

$5 \times 200 = \underline{}$

$5 \times 2{,}000 = \underline{}$

The product of $5 \times 2{,}000$ is $\underline{}$.

On Your Own

a. $6 \times 5 = \underline{}$

$6 \times 50 = \underline{}$

$6 \times 500 = \underline{}$

$6 \times 5{,}000 = \underline{}$

b. $5 \times 9 = \underline{}$

$5 \times 90 = \underline{}$

$5 \times 900 = \underline{}$

$5 \times 9{,}000 = \underline{}$

Write About It

What pattern can you use to solve any of these problems: 500×3, 8×700, 20×20?

Objective 9.1: Multiply multiples of 10.

Lesson 9-1 | **Multiply by Multiples of 10** | **B Understand It**

Example 1

30 × 400 = ?
Basic fact:

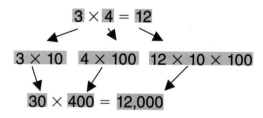

3 is multiplied by 10 to make 30.
4 is multiplied by 100 to make 400.
So multiply the product by 10 and by 100, or by 1,000.

Practice 1

200 × 700 = ?

Basic fact: 2 × 7 = _____

2 is multiplied by _____ to make 200.

7 is multiplied by _____ to make 700.

So multiply the product by _____ and by _____, or by _____.

200 × 700 = _____

Example 2

400 × 500 = ?

Basic fact: 4 × 5 = 20

Multiply by 100 and by 100, or by 10,000.

400 × 500 = 200,000

Practice 2

50 × 600 = ?

Basic fact: _____ × _____ = _____

Multiply by _____ and by _____, or by 1,000.

50 × 600 = _____

On Your Own

a. 300 × 70 = _____

b. 500 × 200 = _____

c. 40 × 4,000 = _____

Write About It

The school stadium can hold 100 students in a row. There are 50 rows. How many students can fit into the stadium? What basic fact can help you solve this problem?

Objective 9.1: Multiply multiples of 10.

Lesson 9-1   **Multiply by Multiples of 10**

1. $9 \times 3 = $ _____

   $9 \times 3\boxed{0} = $ _____

   $9 \times 3\boxed{0}\boxed{0} = $ _____

   $9 \times 3,\boxed{0}\boxed{0}\boxed{0} = $ _____

2. $4 \times 5,000 = ?$

   $4 \times 5 = $ _____

   $4 \times 50 = $ _____

   $4 \times 500 = $ _____

   $4 \times 5,000 = $ _____

   The product of $4 \times 5,000$ is _____.

3. $700 \times 80 = ?$

   Basic fact: $7 \times 8 = $ _____

   Multiply the product by _____.

   $700 \times 80 = $ _____

4. Solve.

   a. $8 \times 4,000 = $ _____

   b. $300 \times 300 = $ _____

   c. $60 \times 900 = $ _____

5. Circle the letter of the correct answer.
   $80 \times 500 = ?$

   A 400   **B** 4,000

   C 40,000   **D** 400,000

6. Circle the letter of the correct answer. Which product will have 4 zeros?

   A 50×500 **B** $20 \times 8,000$

   C 700×50 **D** 500×200

7. Circle the letter of the correct answer. Which product will have 5 zeros?

   A 400×500 **B** $800 \times 5,000$

   C 300×60 **D** $90,000 \times 2$

8. Isaiah is writing a 6-page essay. Each sheet of paper must have 400 words. How many words will be in Isaiah's paper? Explain.

Objective 9.1: Multiply multiples of 10.

Lesson 9-2 — Estimate Products

Model It

Words to Know **Rounding** is changing the digits in a number to make a similar but easier to use number.
An **estimate** is a number close to the actual answer.

Activity 1

Use a number line to estimate 3 × 52.

Round 52 to the greatest place.
52 is closer to 50 than 60.

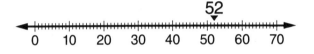

3 × 52 → 3 × 50 = 150

So, 3 × 52 is about 150.

Practice 1

Use a number line to estimate 6 × 37.

Round 37 to the greatest place.
37 rounds to _____.

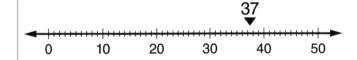

6 × 37 → 6 × _____ = _____

So, 6 × 37 is about _____.

Activity 2

Estimate 24 × 28.

Round 24 to the greatest place.
24 rounds to 20.

Round 28 to the greatest place.
28 rounds to 30.

24 × 28 → 20 × 30 = 600

So, 24 × 28 is about 600.

Practice 2

Estimate 15 × 61.

15 rounds to _____.

61 rounds to _____.

15 × 61 → _____ × _____ = _____

So, 15 × 61 is about _____.

On Your Own

Use a number line to estimate.
  a. 8 × 35

   8 × _____ = _____

  b. 67 × 41

   _____ × _____ = _____

Write About It

Why do you think you round factors to the greatest place?

Objective 9.2: Estimate products by rounding factors and using mental math techniques.

Lesson 9-2 Estimate Products

B Understand It

Example 1

Estimate 183 × 5.

Round 183 to the nearest hundred.
Underline the digit in the hundreds place.
Circle the digit to its right.
If the circled digit is greater than or equal to 5, the underlined digit is increased by 1.

   1⑧3   8 > 5   Change 1 to 2.

183 × 5 → 200 × 5 = 1,000

So, 183 × 5 is about 1,000.

Practice 1

Use rounding rules to estimate 246 × 9.

Round 246 to the nearest _____.

Underline the _____; circle the _____.

246 rounds to _____.

246 × 9 → _____ × 9 = _____

So, 246 × 9 is about _____.

Example 2

Estimate 45 × 239.

Round 45 to the nearest ten.
45 rounds up to 50.

Round 239 to the nearest hundred.
239 rounds down to 200.

Recall the basic fact: 5 × 2 = 10
Use mental math: 50 × 200 = 10,000

So, 45 × 239 is about 10,000.

Practice 2

Estimate 62 × 376.

Round 62 to the nearest _____.

62 rounds _____ to _____.

Round 376 to the nearest _____.

376 rounds _____ to _____.

Recall the basic fact: 6 × 4 = _____

Use mental math: 60 × 400 = _____

So, 62 × 376 is about _____.

On Your Own

Estimate.

   a. 8 × 3,542 → 8 × _____ = _____

   b. 67 × 41 _____ × _____ = _____

Write About It

Why do you say "about" when estimating?

Objective 9.2: Estimate products by rounding factors and using mental math techniques.

Lesson 9-2 Estimate Products

1. Estimate 28 × 7.

   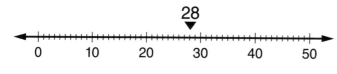

   28 rounds to _____.

   28 × 7 → _____ × 7 = _____

   So, 28 × 7 is about _____.

2. Estimate 56 × 82 using mental math.

   56 rounds to _____.

   82 rounds to _____.

   56 × 82 → _____ × _____ = _____

   So, 56 × 82 is about _____.

3. Estimate 135 × 45 by rounding to the nearest hundred.

   135 × 45 → _____

4. Match each expression with its estimated product.

   18 × 41   1,600

   792 × 2   1,200

   37 × 26   800

   6 × 143   600

5. Estimate.

   a. 84 × 5

   _____ × 5 = _____

   b. 67 × 318

   _____ × _____ = _____

6. 85 × 29 is about

   A 1,600   B 1,800

   C 2,400   D 2,700

7. When you round each factor to the greatest place, which expression has an estimated product of 2,000?

   A 42 × 45   B 56 × 44

   C 376 × 49   D 1,805 × 12

8. Write the steps to estimate 63 × 49.

Objective 9.2: Estimate products by rounding factors and using mental math techniques.

Lesson 9-3 — Multiply: Four Digits by One Digit

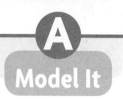

Activity 1

Use base ten blocks to find 213 × 2.

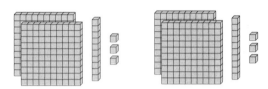

There are 4 hundreds.
There are 2 tens.
There are 6 ones.
213 × 2 = 426

Practice 1

Use base ten blocks to find 322 × 2.

How many hundreds? _____ Tens? _____

Ones? _____ 322 × 2 = _____

Activity 2

Elia and Jon each sold 136 raffle tickets. How many did they sell in all? Use base ten blocks.

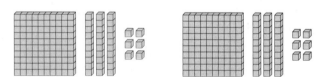

There are 2 hundreds, 6 tens, and 12 ones.
Rename 12 ones as 1 ten and 2 ones.

136 × 2 = 272

Practice 2

Kevin used 325 tiles in each of 2 different designs. How many tiles did he use in all?

Use base ten blocks: _____ tiles

On Your Own

Find 123 × 3.

123 × 3 = _____

Write About It

Tanya had 12 ones, 16 tens, 8 hundreds, and 4 thousands. She renamed twice to get 4,972. Explain how she renamed.

Objective 9.3: Solve simple problems involving multiplication of multidigit numbers by one-digit numbers.

Lesson 9-3 — **Multiply: Four Digits by One Digit** — **B Understand It**

Example 1

Multiply 1,324 × 4.
Use partial products.

1,324 = 1,000 + 300 + 20 + 4
4 × 1,000 = 4,000
4 × 300 = 1,200
4 ×   20 =   80
4 ×   4 =   16
              5,296

1,324 × 4 = 5,296

Practice 1

Multiply 2,153 × 4.
Use partial products.

2,153 = _____

4 × 2,000 = 8,000
4 × _____ = _____
4 × _____ = _____
4 × _____ = _____

2,153 × 4 = _____

Example 2

Tron rode his bicycle 28 miles a day for a week. How far did he ride in that week?

There are 7 days in a week. So multiply 7 × 28.
Write in expanded form.
28 = 20 + 8
7 × 20 = 140
7 × 8 = 56
This can be written in standard form.

```
   5
  28
×  7
 196
```

56 is 5 tens, 6 ones, so 5 is written above the 2 tens in 28 and added to 14 tens.

Practice 2

Cindy can read 36 pages in 1 hour. How many pages can she read in 5 hours?

```
  36
× 5
```

She can read _____ pages.

On Your Own

Multiply 237 × 4 using standard form. Show your work.

```
  237
×   4
```

Write About It

How would you use partial products to solve 5,682 × 5?

Objective 9.3: Solve simple problems involving multiplication of multidigit numbers by one-digit numbers.

Lesson 9-3: Multiply: Four Digits by One Digit

Try It

1. Write the multiplication problem shown.

   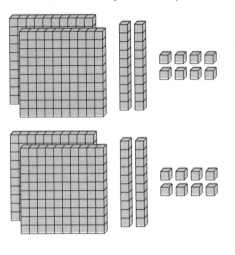

   _____

2. Multiply. Use base ten blocks.

   a. 23 × 3 = _____

   b. 430 × 2 = _____

3. Multiply. Use partial products.

   a. 89 × 8 = _____

   b. 7 × 42 = _____

4. What is 346 × 5? Circle the letter of the correct answer.

   A 1,500   **B** 1,700

   C 1,730   **D** 1,740

5. Solve 34 × 2. _____

   Use the answer to find 340 × 2. _____

   Find 3,400 × 2. _____

6. On a Monday, Mr. Rojas teaches 4 sessions of history classes. Each class has 18 students. How many students attend Mr. Rojas' classes on Monday?

   _____

   _____

7. Rand walked 6 miles each day in July. He decided he had walked 1,806 miles. Is his answer reasonable? Why or why not?

   _____

8. Explain two different ways to find the product of 312 × 3.

   _____

   _____

Objective 9.3: Solve simple problems involving multiplication of multidigit numbers by one-digit numbers.

Lesson 9-4: Choose a Method and Multiply

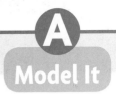
Model It

Words to Know Doing **mental math** means solving math problems in your head.

Activity 1

87 × 64 = ?

Estimate.   87 × 64
             ↓   ↓
            90 × 60 = 5,400

Then calculate.   87 × 64 = 5,568

Compare 5,568 to your estimate.

Practice 1

93 × 79 = ?

Estimate.   93 × 79
             ↓   ↓
           _____ × _____ = _____

Then calculate. 93 × 79 = _____
Is your estimate close to your exact answer? _____

Activity 2

The fifth grade is designing a float for the parade. There are 87 students in fifth grade. If each student makes 23 paper flowers, how many total flowers will they have to decorate the float?

Estimate.   87 × 23 → 90 × 20 = 1,800
Then calculate.   87 × 23 = 2,001

Compare 2,001 to your estimate.
The estimate is close to the exact answer.
They will have 2,001 paper flowers.

Practice 2

The art teacher bought 18 packs of paper. Each pack contained 25 sheets. How many sheets of paper did the teacher buy?

Estimate: _____ × _____ = _____

Then calculate. _____
Is your estimate close to your exact answer? _____

The teacher bought _____ sheets.

On Your Own

Estimate the product of 73 × 82.

73 × 83 → _____ × _____ = _____

Then find the exact answer.

173 × 82 = _____

Write About It

Can you use mental math to solve any of the problems on this page? Why or why not?

Objective 9.4: Solve problems by choosing between using estimation, mental math, or paper and pencil.

Lesson 9-4 — Choose a Method and Multiply

B Understand It

Words to Know — To use the **partial-products method,** multiply the value of each digit in one factor by the value of each digit in the other factor. Add the partial products.

Example 1

Tickets to the amusement park cost $36 each. What is the cost for a family of 5?

Write a problem. $36 × 5
Is there a way to solve mentally?
There are only two digits to multiply by 5.
Use partial products to solve mentally.

$$36 = 30 + 6$$
Multiply 6 × 5 mentally: 30.
Multiply 30 × 5 mentally: 150.
Add the partial products: 30 + 150 = 180
The cost is $180.

Practice 1

Carter plays the piano 45 minutes each day. How much time does he play each week?

Write an problem. _____

Is there a way to solve mentally?
Solve.

Carter plays _____ minutes each week.

Example 2

Brett pays $38.95 a month for phone service. How much does he spend in two months?
You need to find an exact answer.
Write a problem. $38.95 × 2
Is there a way to solve mentally?
You know the twos, but you must find and add 4 partial products. Use pencil and paper.

```
   1 1 1
$ 38.95
   ×  2
───────
$ 77.90
```
Brett spends $77.90 in two months.

Practice 2

In a video game, a player scores 500 points for completing a level. Josh completed 4 levels. What is his score?
Choose a method.
    mental math   paper and pencil   estimation
Explain your choice.

Solve.

Josh's score is _____ points.

On Your Own

Lee earns $47 a week delivering newspapers. Much does Lee earn in a year?

Lee earns about _____ a year.

Write About It

How do you decide which method to use?

Objective 9.4: Solve problems by choosing between estimation, mental math, or pencil and paper.

Lesson 9-4 — Choose a Method and Multiply

1. 71 × 86 = ?

Estimate. 71 × 86
↓   ↓
_____ × _____ = _____

Calculate.

Check.

The product of 71 × 86 is _____.

2. The drama club sold 429 tickets for $9 each. How much did the drama club make in ticket sales?

The drama club made _____.

3. Mr. Jones drove for 6 hours at 58 miles per hour. How many miles did Mr. Jones drive?

Choose a method.

mental math   paper and pencil   estimate

Explain your choice.

Mr. Jones drove _____ miles.

4. Each camper uses 36 beads to make a necklace. There are 41 campers. About how many beads are needed?

Choose a method.

mental math   paper and pencil   estimate

Explain your choice.

About _____ beads are needed.

5. A photo album has 72 pages. Each page holds 6 photos. How many photos are in the album?

A 412   B 420

C 432   D 700

6. Which problem is easier to solve mentally: 203 × 4 or 7 × 36? Explain your choice.

7. Write a multiplication problem that can be solved using paper and pencil. Show the solution.

8. Write a multiplication problem that you can solve mentally. Show the solution.

Objective 9.4: Solve problems by choosing between using estimation, mental math, or pencil and paper.

Topic 9: Use Multiplication to Compute

Topic Summary

Choose the correct answer. Explain how you decided.

1. A class of 5th graders is going to the Science Center. If there are 134 students and tickets cost $8.75 each, **about** how much will it cost for all the students to go to the Science Center?

   A $1,000.00

   B $1,172.50

   C $1,300.00

   D $1,400.00

   _____
   _____
   _____

2. A scrapbook has 45 pages. Each page holds 8 pictures. How many pictures are in the scrapbook?

   A 53

   B 320

   C 360

   D 450

   _____
   _____
   _____

Objective: Review computations with multiplication.

Topic 9: Use Multiplication to Compute

Mixed Review

1. Complete the pattern.

   a. $6 \times 7 =$ _____

   b. $60 \times 7 =$ _____

   c. $60 \times 70 =$ _____

   d. $600 \times 7 =$ _____

   e. $600 \times 70 =$ _____

   Volume 4, Lesson 9-1

2. Write the place of the underlined digit in 6,8̲71.

   _____

   Volume 1, Lesson 3-1

3. Dylan added $69 + 25$ and got 84. Is Dylan correct? Why or why not?

   _____

   _____

   _____

   Volume 1, Lesson 1-2

4. Willis has 56 football cards. He gave some to his brother. Willis now has 37 cards. Choose the number sentence that shows how many cards Willis gave his brother.

   A $56 + 37 = 93$

   B $37 + 56 = 93$

   C $37 - 56 = 19$

   D $56 - 37 = 19$

   Volume 2, Lesson 4-1

5. What is eight million, six hundred thirty-eight thousand, two hundred five in standard form?

   A 8,683,250   B 8,638,250

   C 8,638,205   D 8,608,205

   Volume 1, Lesson 3-3

6. Find each quotient.

   a. $72 \div 9 =$ _____

   b. $12 \div 4 =$ _____

   c. $49 \div 7 =$ _____

   Volume 2, Lesson 5-1

7. Multiply.

   $24 \times 26 =$ _____

   Volume 2, Lesson 7-4

8. Write the first six multiples of 8.

   _____

   Volume 1, Lesson 4-3

Objective: Maintain skills and concepts.

Topic 10: Use Division to Compute

Topic Introduction

Complete with teacher help if needed.

1. Divide.

   a. 28 ÷ 4 = _____

   b. Complete the pattern.

   280 ÷ 4 = _____

   280 ÷ 40 = _____

   2,800 ÷ 400 = _____

Objective 10.1: Solve problems involving dividing multiples of 10.

2. Use compatible numbers to estimate 3,786 ÷ 5.

   a. Is 37 ÷ 5 a basic fact? _____

   b. What number is close to 37 and a multiple of 5? _____

   c. 3,786 ÷ 5 ⟶ _____ ÷ 5 = _____

Objective 10.2: Estimate quotients by rounding factors and using mental math techniques.

3. Divide.

   a. Use base ten blocks to divide 112 ÷ 8.

   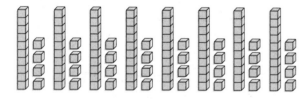

   b. 112 ÷ 8 = _____

Objective 10.3: Solve simple problems involving division of multidigit numbers by 1-digit numbers with and without remainders.

4. A physical education teacher purchased 486 jump ropes. If 9 jump ropes are packaged in a box, how many boxes will it take to ship all of the jump ropes?

   a. Will you multiply or divide? _____

   b. 486 ÷ 9 = _____

Objective 10.4: Solve problems by multiplying or dividing.

Lesson 10-1 — **Divide Multiples of 10**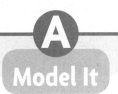

Words to Know dividend → 40 ÷ 8 = 5 ← quotient
                              ↑
                          divisor

Activity 1

Find the quotients.

6 ÷ 3 = 2
60 ÷ 3 = 20
600 ÷ 3 = 200
6000 ÷ 3 = 2000

Practice 1

Find the quotients.

8 ÷ 2 = ___
80 ÷ 2 = ___
800 ÷ 2 = ___
8000 ÷ 2 = ___

Activity 2

Find the quotients.

60 ÷ 30 = 2
600 ÷ 30 = 20
6000 ÷ 30 = 200

Practice 2

Find the quotients.

80 ÷ 20 = ___
800 ÷ 20 = ___
8000 ÷ 20 = ___

On Your Own

Solve.

a. 40,000 ÷ 2,000 = ___

b. 180,000 ÷ 600 = ___

c. 900 ÷ 30 = ___

Write About It

Explain how to solve 36,000 ÷ 600.

Objective 10.1: Solve problems dividing multiples of 10, 100, and 1,000.

Volume 4   17   Level E

Lesson 10-1 **Divide Multiples of 10**   **B Understand It**

Example 1

Find each quotient.

4,200 ÷ 7 = 600
4,200 ÷ 70 = 60
4,200 ÷ 700 = 6

Basic fact:
42 ÷ 7 = 6

Practice 1

Find each quotient.

3,600 ÷ 9 = _____
3,600 ÷ 90 = _____
3,600 ÷ 900 = _____

Basic fact:
36 ÷ 9 = _____

Example 2

Find the quotients.

4,000 ÷ 8 = 500
4,000 ÷ 80 = 50
4,000 ÷ 800 = 5

Practice 2

Find the quotients.

2,000 ÷ 4 = _____
2,000 ÷ 40 = _____
2,000 ÷ 400 = _____

On Your Own

Find each quotient. Give the basic fact for each one.

a. 2,800 ÷ 4 = _____

   basic fact: _____

b. 7,200 ÷ 80 = _____

   basic fact: _____

c. 3,000 ÷ 500 = _____

   basic fact: _____

Write About It

What is the basic fact in 30,000 ÷ 6? 3 ÷ 6, 30 ÷ 6, or 300 ÷ 6? Explain. Then solve 30,000 ÷ 6.

Objective 10.1: Solve problems dividing multiples of 10, 100, and 1,000.

Lesson 10-1 — **Divide Multiples of 10**

Try It

1. Find the quotients.

   4 ÷ 2 = _____

   40 ÷ 2 = _____

   400 ÷ 2 = _____

   4000 ÷ 2 = _____

2. Find each quotient.

   6,300 ÷ 9 = _____

   6,300 ÷ 90 = _____

   6,300 ÷ 900 = _____

   Basic fact:
   63 ÷ 9 = _____

3. What is 2,000 ÷ 5? Circle the letter of the correct answer.

   A 4   B 40

   C 400   D 4,000

4. A store manager decided to divide 500 marbles into 10 bags to sell. How many marbles were in each bag?

   500 ÷ 10 = _____

   There were _____ marbles in each bag.

5. Solve. Write the basic fact in each one.

   a. 490 ÷ 7 = _____

      basic fact _____

   b. 3,200 ÷ 400 = _____

      basic fact _____

   c. 45,000 ÷ 500 = _____

      basic fact _____

6. Which of the following is **not** equal to 5?

   A 300 ÷ 60   B 3,000 ÷ 600

   C 30 ÷ 6   D 300 ÷ 600

7. Match each division expression with its quotient.

   1,200 ÷ 400   30

   900 ÷ 30   300

   2,400 ÷ 8   3

8. Explain how to divide 1,000 ÷ 5.

Objective 10.1: Solve problems dividing multiples of 10, 100, and 1,000.

Lesson 10-2: Estimate Quotients

Words to Know An **estimate** is a number close to the exact answer.

Activity 1

Estimate 7,615 ÷ 4.

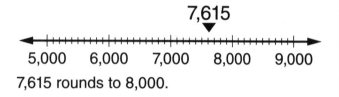

7,615 rounds to 8,000.

Think: 8,000 ÷ 4 = 2,000.

The quotient of 7,615 ÷ 4 is about 2,000.

Practice 1

Estimate 647 ÷ 3.

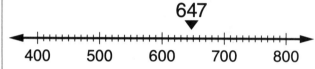

647 rounds to _____.

Think: _____ ÷ 3 = _____.

The quotient of 647 ÷ 3 is about _____.

Activity 2

Estimate 2,040 ÷ 43.

2,040 rounds to 2,000.
43 rounds to 40.

2,040 ÷ 43 → 2,000 ÷ 40 = 50

So 2,040 ÷ 43 ≈ 50.

Practice 2

Estimate 5,712 ÷ 31.

5,712 rounds to _____.
31 rounds to _____.

5,712 ÷ 31 → _____ ÷ _____ = _____

So 5,712 ÷ 31 ≈ _____.

On Your Own

Estimate each quotient.

a. 910 ÷ 328

_____ ÷ _____ = _____

b. 8,146 ÷ 217

_____ ÷ _____ = _____

Write About It

Write out the steps that you would use when estimating a quotient.

Objective 10.2: Estimate quotients by rounding numbers and using mental math techniques.

Lesson 10-2: Estimate Quotients

Understand It

Words to Know Compatible numbers are close to the actual numbers but easier to work with mentally.

Example 1

Use compatible numbers to estimate.

$$4,415 \div 7$$

$44 \div 7$ is **not** a basic fact.
42 is close to 44 and is a multiple of 7.

$$4,415 \div 7$$
$$\downarrow \quad \downarrow$$
$$4,200 \div 7 = 600$$

$4,415 \div 7 =$ about 600

Practice 1

Use compatible numbers to estimate.

$$3,110 \div 8$$

Is $31 \div 8$ a basic fact? _____
What number is close to 31 **and** a multiple of 8? _____

$$3,110 \div 8$$

_____ $\div 8 =$ _____

$3,110 \div 8 =$ about _____

Example 2

Estimate. Round the divisor first, and then choose a compatible dividend.

$5,237 \div 612$
$\downarrow$
$5,237 \div 600$
$\downarrow$
$5,400 \div 600 = 9$

Think: $54 \div 6$ is a basic fact.
5,400 is close to 5,237.
Use 5,400.

So $5,237 \div 612 \approx 9$.

Practice 2

Estimate $1,476 \div 35$.

35 rounds to _____.
What basic fact can you use to find a compatible number for 1,476? _____

$$1,476 \div 35$$

_____ $\div$ _____ = _____

So $1,476 \div 35 \approx$ _____.

On Your Own

Use compatible numbers to estimate.
a. $3,321 \div 53$
   $\downarrow \quad \downarrow$
   _____ $\div$ _____ = _____

b. $836 \div 94$
   $\downarrow \quad \downarrow$
   _____ $\div$ _____ = _____

Write About It

It is **not** helpful to round some dividends to the greatest place to estimate. Why?

Objective 10.2: Estimate quotients by rounding numbers and using mental math techniques.

Lesson 10-2 — **Estimate Quotients**

1. Round to estimate 429 ÷ 5.

   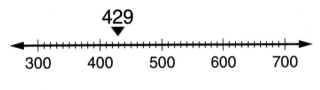

   429 rounds to _____.

   Think: _____ ÷ 5 = _____

   429 ÷ 5 = about _____

2. Use compatible numbers to estimate.

   3,408 ÷ 9

   What basic fact can you use to round the dividend? _____

   3,408 ÷ 9
     ↓   ↓
   _____ ÷ 9 = _____

   3,408 ÷ 9 ≈ _____

3. For the **best** estimate of 4,125 ÷ 60, which number would you choose as a compatible number for the dividend? Circle the letter of the correct answer.

   A 6,000   **B** 4,000

   C 4,200   **D** 4,800

4. Round to estimate 7,833 ÷ 421.

   7,833 rounds to _____.
   421 rounds to _____.

   So, 7,833 ÷ 421 ≈ _____

5. Use compatible numbers to estimate.
   5,800 ÷ 655

   _____ ÷ _____ = _____

6. The tennis coach collected 174 tennis balls. If 3 balls fit in one container, about how many containers will the coach need for all the balls?

   _____ ÷ _____ = _____

7. Write a division expression with an estimated quotient of 50.

8. Marcie and Collin both use compatible numbers to estimate the quotient of 3,020 ÷ 741. Marcie rounds the problem to 3,500 ÷ 700. Collin rounds the problem to 2,800 ÷ 700. Whose estimate will be closest to the exact answer? How do you know?

Objective 10.2: Estimate quotients by rounding numbers and using mental math techniques.

Lesson 10-3: Divide by 1-Digit Numbers

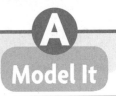
Model It

Activity 1

Use base ten blocks to divide 693 ÷ 3.

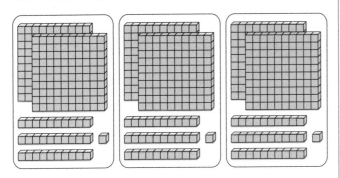

693 ÷ 3 = 231

Practice 1

Use base ten blocks to divide 428 ÷ 2.

428 ÷ 2 = _____

Activity 2

Use base ten blocks to divide 87 ÷ 4.

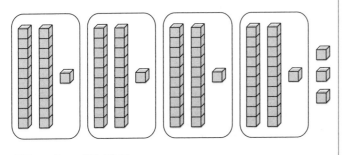

87 ÷ 4 = 21 R 3

Practice 2

Use base ten blocks to divide 38 ÷ 3.

38 ÷ 3 = _____

On Your Own

Use base ten blocks to divide 645 ÷ 2.

645 ÷ 2 = _____

Write About It

When will a division problem have a remainder?

Objective 10.3: Solve simple problems involving division of multidigit numbers by one-digit numbers with and without remainders.

Lesson 10-3 Divide by 1-Digit Numbers

B Understand It

Example 1

Use paper and pencil to find 175 ÷ 7. Check your answer.

175 ÷ 7 = 25

```
      13 R5
   7)96
      7
      26
      21
       5
```

Practice 1

Use paper and pencil to find 276 ÷ 4. Check your answer.

276 ÷ 4 = _____

Example 2

Use paper and pencil to find 456 ÷ 12. Check your answer.

456 ÷ 12 = 38

```
       38
   12)456
      36
       96
       96
        0
```

Practice 2

Use paper and pencil to find 627 ÷ 11. Check your answer.

627 ÷ 11 = _____

On Your Own

Use paper and pencil to find 3,740 ÷ 5. Check your answer.

3,740 ÷ 5 = _____

Write About It

How can an estimate using mental math help you when using pencil and paper to divide?

Objective 10.3: Solve simple problems involving division of multidigit numbers by one-digit numbers with and without remainders.

Lesson 10-3: Divide by 1-Digit Numbers

Try It

1. Use paper and pencil to find 84 ÷ 3.

2. Use paper and pencil to find 2,637 ÷ 9.

3. Which division problem is modeled?

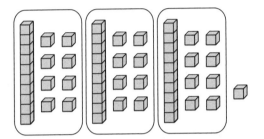

A 55 ÷ 3 **B** 55 ÷ 18

C 55 ÷ 4 **D** 55 ÷ 1

4. Which problem has a remainder?

A 814 ÷ 4   **B** 475 ÷ 5

C 447 ÷ 3   **D** 142 ÷ 2

5. Maya has 37 pictures of her classmates. She wants to arrange them into groups of 6. How many groups will she make? How many pictures, if any, will be left over?

6. Which paper and pencil division is done correctly for 746 ÷ 2?

```
    A   323      B   328      C   373      D   473
      2)746        2)746        2)746        2)746
        6            6            6            8
        4            6            14           14
        4            5            14           14
        6            4            6            6
        6            16           6            6
                     16
```

7. Use base ten blocks to divide 181 ÷ 5.

181 ÷ 5 = _____

8. Mario puts his books on 8 shelves in his bookcase. If he has 97 books, how many books will he put on each shelf?

Objective 10.3: Solve simple problems involving division of multidigit numbers by one-digit numbers with and without remainders

Lesson 10-4 **Choose a Method for Division**

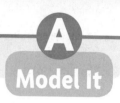

Activity 1

Estimate $714 \div 3$.

Think: 714 rounded to the nearest hundred is 700.

Think: $3 \times 200 = 600$ and $3 \times 300 = 900$, so the answer is between 200 and 300.

$714 \div 3 \approx 250$

Practice 1

Estimate $2,358 \div 6$.

2,358 rounded to the nearest hundred is _____.

$6 \times $ _____ $= 24$, so $6 \times$ _____ $= 2,400$

$2,358 \div 6 \approx$ _____

Activity 2

Estimate $1,312 \div 32$.

Think: 1,312 rounded to the nearest hundred is 1,300, and 32 rounded to the nearest ten is 30.

Think: There are about 3 groups of 30 in one hundred. How many groups of 30 are in 13 hundreds?

3 groups of 30 $\times$ **13** hundreds $= 39$

$1,312 \div 32 \approx 39$

Practice 2

Estimate $2,698 \div 21$.

2,698 rounded to the nearest hundred is _____ , and 21 rounded to the nearest ten is _____.

There are _____ groups of 20 in 1 hundred. How many groups of 20 are in _____ hundreds?

_____ groups of 20 $\times$ _____ hundreds = _____

$2,698 \div 22 \approx$ _____

On Your Own

Use mental math to find $4,800 \div 60$.

$4,800 \div 60 =$ _____

Write About It

Does mental math always provide just an estimate or can it give an exact answer? Explain.

Objective 10.4: Solve division problems by choosing between using mental math, estimation, or pencil and paper to find the quotients.

Lesson 10-4: Choose a Method for Division

B Understand It

Example 1

The store clerk needs to put 175 candies in 7 gift bags. How many candies should be placed in each bag?
Use paper and pencil to find 175 ÷ 7.

```
      25
  7)175
     14
     35
     35
      0
```

175 ÷ 7 = **25 candies**

Practice 1

Henry has 4 vases and 276 flowers. How many flowers should go in each vase?
Use paper and pencil to find 276 ÷ 4.

276 ÷ 4 = _____

Example 2

Marcia read 523 pages in 12 hours. About how many pages did she read per hour?

523 rounded to the nearest ten is **520**, and 12 rounded to the nearest ten is **10**.

10 × **52** = 520

523 ÷ 12 ≅ **52**

Marcia read about **52 pages per hour**.

Practice 2

In 32 hours, a machine produced 586 toys. About how many toys did the machine produce per hour?

586 rounded to the nearest hundred is _____, and 32 rounded to the nearest ten is _____.

3 × _____ = 6, so 30 × _____ = 600

The machine produced about _____.

On Your Own

Use mental math to find 3,500 ÷ 50.

_____

Write About It

When rounding numbers in division problems, how do we decide whether to round to tens, hundreds, or thousands?

Objective 10.4: Solve division problems by choosing between using mental math, estimation, or pencil and paper to find the quotients.

Lesson 10-4: Choose a Method for Division

Try It

1. Estimate $2{,}510 \div 39$.

   $2{,}510 \div 39 \approx$ _____

2. Which calculation will provide an estimation that is closest to the exact answer?

   A $3{,}000 \div 50 = 60$ **B** $3{,}000 \div 60 = 50$

   C $2{,}800 \div 70 = 40$ **D** $2{,}400 \div 60 = 40$

3. After working for 90 hours, Harrison earned $4,500. How much did he earn per hour?

   _____

4. Toni opened 13 bags of snacks and counted 166 total snacks. About how many were in each bag?

   _____

5. Which is greater: $137 \div 8$, or $232 \div 12$?

   _____

   How did you decide?

6. A series of 30 books has a total of 3,250 pages. If each book is the same length, about how many pages are in each book?

   _____

7. Find an exact answer for $2{,}001 \div 23$.

   $2{,}001 \div 23 =$ _____

   Use estimation to show that your answer is reasonable.

8. Tameca and Selena want to build a tree fort. They need 153 feet of lumber. If the lumber comes in 8-foot lengths and 9-foot lengths, how many pieces of lumber will they need if they use:

   a. 8-foot lengths of lumber? _____

   b. 9-foot lengths of lumber? _____

Objective 10.4: Solve division problems by choosing between using mental math, estimation, or pencil and paper to find the quotients.

Lesson 10-5: Multiplication and Division

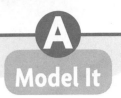
Model It

Activity 1

A cook at a diner prepares 3-egg omelets. On an average day, 78 people order the omelet. About how many eggs does the cook use each day?

78 rounds to 80.
$80 \times 3 = 240$

The cook uses about 240 eggs each day.

Practice 1

One bag of popcorn makes a snack for 8 people. About how many people can have a snack using 18 bags of popcorn?

18 rounds to _____.

$20 \times 8 =$ _____

The popcorn can feed about _____ people.

Activity 2

Movie packages can be purchased from a satellite television company. Each package has 12 channels. How many movie channels are purchased with 3 packages?

$12 \times 3 = 36$

There are 36 movie channels in 3 packages of 12 movie channels.

Practice 2

There are 24 pictures on a roll of film. Janie used 4 rolls of this film to take pictures of her vacation. How many pictures did Janie take all together?

$24 \times 4 =$ _____

Janie took _____ pictures.

On Your Own

A flooring company is placing tile on the floor of a room. 17 tiles are put in each row. 21 rows of tile are needed. About how many tiles will be used?

17 rounds to _____.

21 rounds to _____.

$20 \times 20 =$ _____

About _____ tiles are needed for the floor.

Write About It

Write a word problem that 2×9 can be used to solve.

Objective 10.5: Solve problems by multiplying or dividing.

Lesson 10-5: Multiplication and Division

Understand It — B

Example 1

A book has 112 pages. 8 pages can be read in an hour. How many hours will be needed to finish reading the book? Use paper and pencil to answer the question.

It will take 14 hours to finish reading the book.

```
      14
   8)112
      8
      32
      32
```

Practice 1

72 students are going on a field trip. 6 students can go in each van. How many vans are needed for the field trip?

There are 72 ÷ _____ vans needed.

72 ÷ 6 = _____

There are _____ vans needed for the trip.

Example 2

A teacher purchased 20 calculators for her classroom. The total cost was $242. What was the approximate cost of each calculator?

242 rounds to 200.
200 ÷ 20 = 10

Each calculator cost about $10.

Practice 2

A case of raisins holds 46 boxes of raisins. Approximately how many cases must be purchased to have 368 boxes of raisins?

368 rounds to _____.
46 rounds to _____.
400 ÷ 50 = _____

About _____ cases of raisins must be purchased.

On Your Own

A family played 3 hours of tennis on Saturday afternoon. If each set was 30 minutes long, how many sets did they play?
(1 hour = 60 minutes)

60 × 3 = _____

180 ÷ 30 = _____

The family plays _____ sets of tennis that afternoon.

Write About It

Write a word problem that 20 ÷ 2 can be used to solve.

Objective 10.5: Solve problems by multiplying or dividing.

Lesson 10-5: Multiplication and Division

1. Scotty earned $135 for mowing lawns one weekend. He worked for 9 hours. How much did he earn per hour?

Scotty earned $_____ per hour.

2. There are 365 days in a year. How many days are there in 2 years?

There are _____ days in 2 years.

3. Which equation could be used to solve the following problem? Circle the letter of the correct answer.

There are 12 eggs in a dozen. How many eggs are there in 4 dozen?

A 12×4 B $12 + 4$

C $12 \div 4$ D $12 - 4$

4. You are given the total number of participants in a tournament and the number of teams. What operation is used to find the number of participants on each team? Circle the letter of the correct answer.

A Addition   B Subtraction

C Multiplication   D Division

5. Which of the following problems could use $24 \div 6$ to solve it? Circle the letter of the correct answer.

A One book costs $24. How much will 6 books cost?

B $24 is spent on 6 books. What is the cost of each book?

C A book costs $24 and is on sale for $6 off. What is the sale price of the book?

D Several books totaling $24 were purchased online. The shipping cost is $6. What is the total cost of the books?

6. Chairs are being placed in a room. There are 36 chairs in each of 22 rows. Write an equation to solve the problem. Explain the chosen operation.

7. Estimate the solution to Exercise 6.

There are approximately _____ chairs in the room.

8. A waiter has 32 drinks to deliver to a large party. He can put 8 drinks on a tray. How many trays will he need?

He will need _____ trays of drinks.

Objective 10.5: Solve problems by multiplying or dividing.

Topic 10: Use Division to Compute

Topic Summary

Choose the letter of the correct answer. Explain how you decided.

1. What is 30,000 ÷ 60?

   A 500

   B 5,000

   C 50,000

   D 500,000

   _____
   _____
   _____
   _____

2. Jason's music player can hold 363 songs. If he can download 3 songs in a minute, how long will it take him to fill his music player?

   A 120

   B 121

   C 366

   D 1,089

   _____
   _____
   _____

Objective: Review computation using division.

Topic 10 — Use Division to Compute

Mixed Review

1. A car dealership sells 87 cars each month. How many cars will they sell in a year?

   A 870   B 957

   C 1,044   D 1,131

   Volume 4, Lesson 9-3

2. At football practice, there are 22 players on the field and 19 more on the sidelines. How many players are at practice?

   A 31   B 41

   C 418   D 51

   Volume 3, Lesson 7-2

3. Order 610, 61, and 601 from least to greatest.

   _____

   Volume 1, Lesson 2-3

4. Amanda estimated 5,683 ÷ 7 as 6,000 ÷ 7. What would be a better estimate?

   _____

   _____

   _____

   Volume 4, Lesson 10-2

5. Write the place of each underlined digit.

   a. 8,0__3__6 _____

   b. 27__4__ _____

   c. __1__,224 _____

   d. 3,__5__08 _____

   Volume 1, Lesson 2-1

6. Find each quotient.

   a. 14 ÷ 7 = _____

   b. 12 ÷ 1 = _____

   c. 36 ÷ 4 = _____

   d. 24 ÷ 3 = _____

   Volume 2, Lesson 6-4

7. Use the make-ten strategy to find the sum 7 + 5.

   _____

   _____

   Volume 2, Lesson 4-3

8. Subtract 572 − 37.

   _____

   Volume 3, Lesson 8-3

Objective: Maintain concepts and skills.

Topic 11: Expressions and Equations

Topic Introduction

Complete with teacher help if needed.

1. The operation modeled below includes an unknown.

   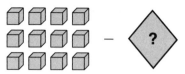

   a. What operation is modeled?

   _____

   b. Which value is known?

   _____

   c. What represents the unknown?

   _____

   Objective 11.2: Write expressions for situations that include an unknown quantity.

2. Evaluate each expression.

   a. $x \div 3$ for $x = 6$

   _____

   b. $b - 1$ for $b = 3$

   _____

   c. $2x$ for $x = 10$

   _____

   Objective 11.2: Write expressions for situations that include an unknown quantity.

3. A set of earrings costs $10.

   a. How would you find the price of 8 sets of earrings? Write an expression.

   _____

   b. How would you find the price of e sets of earrings? Write an expression.

   _____

   Objective 11.1: Record the rule for a pattern as an expression.

4. Kelsey has two boxes of baseball cards. One box has 30 cards in it. Kelsey is not sure how many cards are in the other box. She knows that she has a total of 95 cards.

   a. Which values does she know?

   _____

   b. What is the total? _____

   c. Write an addition equation to represent this situation. Use a letter for the unknown amount.

   _____

   Objective 11.3: Write equations for word problems that include an unknown quantity.

Lesson 11-1: Write Expressions for Patterns

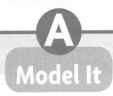
Model It

Words to Know An **unknown** is a letter or a symbol that stands for a number.
An **algebraic expression** is a mathematical phrase with at least one unknown and one operation.

Activity 1

■ + ○○○

The expression is ■ + 3.

What is the value if ■ = 6?

3 + _____ = _____

Practice 1

○○
○○   △

Write the expression.

_____

What is the value if △ = 2?

4 + _____ = _____

Activity 2

Harlon earns $10 for each hour that he works. How much will he earn next week?

His earnings can be found by multiplying his hourly wage by the number of hours he works.

For **1** hour, he earns 10 × 1 or 10(**1**).

For **3** hours, he earns 10 × 3 or 10(**3**).

For h hours, he earns 10 × h or 10(h) or 10h.

Harlon will earn 10h$ next week.

Practice 2

A book sells for $9. Write an expression to show the price for each **purchase**.

4 books _____

10 books _____

300 books _____

Write an algebraic expression to show the price for b books.

_____

On Your Own

There are 8 boys and some girls in a class. Write an expression to show the number of students in the class. Find the value if there are 6 girls.

_____

Write About It

Bags of peanuts cost $3 each. Find the cost of 2, 3, and 4 bags of peanuts. Then write an expression to show the price for p bags of peanuts.

_____

Objective 11.1: Record the rule for a pattern as an expression.

Lesson 11-1: Write Expressions for Patterns

B Understand It

Example 1

Each salad includes two tomatoes. Colin made a table to show the number of tomatoes in different numbers of salads. The Input is the number of salads, and the Output is the number of tomatoes.

Input	2	3	4	5	★
Output	4	6	8	10	2★

The rule for the table is *Multiply by 2.* Another way to write the rule is to use an unknown: 2 × ★ or 2★.

Practice 1

Complete the table to find the cost of different numbers of movie tickets. The Input is the number of tickets, and the Output is the cost of those tickets in dollars.

Input	5	6	7	8
Output	25	30	35	

Write the rule in words. Then write the rule using an unknown.

_____

Example 2

There are 3 feet in a yard.

Yard	1	2	3	y
Feet	3	6	9	3y

The rule for the table is Multiply by 3. The last column of the table shows the rule as an expression: If there are y yards, there are $3y$ feet.

Practice 2

The perimeter of a regular octogon is 8 times the length of one of its sides.

Fill in the blanks in the table.

Reg. Octogon Side Length	1	2	3	4	o
Perimeter	8	16	24		

On Your Own

There are 4 quarters in a dollar.

Fill in the blanks in the table.

Dollar	1	2	3	4	5
Quarters	4	8	12		

Write About It

Look at the table in On Your Own. Write the rule in words. Then write the rule using an unknown.

Objective 11.1: Record the rule for a pattern as an expression.

Lesson 11-1 — Write Expressions for Patterns

Try It

1. Write an algebraic expression for each of the following.

   a. $7 + x$

   _____

   b. $y - 10$

   _____

2. For each table, write the rule in words and as an algebraic expression. Use n for the unknown. Complete each table.

   a.

Input	1	2	3	4
Output	2	4	6	

   Rule: _____

   b.

Input	10	20	30	
Output	14	24	34	44

   Rule: _____

3. Magazines sell for $3 each. Write an expression to show the price for each purchase.

   2 magazines _____

   4 magazines _____

   m magazines _____

4. There were 10 ducks in a pond. Then some of them flew away. Write an algebraic expression to show the number of ducks that are left. _____

5. $5 + \bigcirc$

   If $\bigcirc = 10$, what is the value? _____

6. Fill in the blanks in the table.

Rectangles	1	2	3	4	r
Sides	4	8			

Objective 11.1: Record the rule for a pattern as an expression.

Lesson 11-2 Write Expressions

Activity 1

At the entrance to a shopping center, the sign for the furniture store is twice as long as the sign for the card shop.

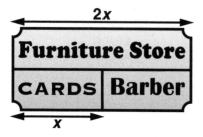

The card shop sign is x feet long. The furniture store sign is twice as long, or $2x$ ft.

Practice 1

Complete the picture. Show that Mee's walk to school is three times as long as Jon's walk.

Write expressions for the length of Jon's walk and the length of Mee's walk. Use the same variable in each expression.

Jon _____

Mee _____

Activity 2

The expression shows the cost of x cartons of oil paints.

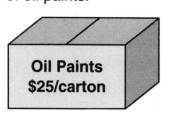

$25x$

How much would 20 cartons of oil paints cost? Substitute 20 for x:
$25(20) = 500$
The paints would cost $500.

Practice 2

Mr. Gomez charges $15 per hour for guitar lessons. Write an algebraic expression for the income from guitar lessons.

income from n guitar lessons: _____

What is Mr. Gomez's income if he gives 5 hours of guitar lessons? Evaluate your expression to solve.

_____

On Your Own

A package of doughnuts costs $3. Write an expression for the cost of d packages of doughnuts. Jim buys 4 packages of doughnuts for a party. How much does he spend?

Write About It

Jim wrote 5 less than a number as $5 - x$. Is his expression correct? Explain.

Objective 11.2: Write expressions for situations that include an unknown quantity.

Lesson 11-2: Write Expressions

Understand It — B

Words to Know
3 **fewer than** a number $x - 3$
5 **greater than** a number $y + 5$

Example 1

Jackie bought T-shirts for her club. She bought 3 fewer white shirts than blue shirts.

If there are x blue shirts, then there are $x - 3$ white shirts.

If Jackie bought 7 blue shirts, how many white shirts did she buy?

Substitute 7 for x: $7 - 3 = 4$.

She bought 4 white shirts.

Practice 1

Paula's locker number is 32 greater than Stuart's locker number.

Write an algebraic expression for each number. Use the variable n in each expression.

Stuart's locker number _____

Paula's locker number _____

What is Paula's locker number if Stuart has locker number 26? _____

Example 2

Write the algebraic expressions.

Word expression	Algebraic expression
7 less than a number b	$b - 7$
the product of a number a and 9	$9a$
50 divided by a number n	$50 \div n$
2 more than a number r	$r + 2$

Practice 2

Write the algebraic expressions.

Word expression	Algebraic expression
6 greater than a number t	_____
a number q divided by 5	_____
the difference of a number k and 3	_____

On Your Own

Evaluate $4m$ for $m = 4$.

_____

Write About It

Write two different word expressions that can be shown by $12 - x$.

Objective 11.2: Write expressions for situations that include an unknown quantity.

Lesson 11-2 **Write Expressions**

Try It

1. The band's T-shirts cost $12. Posters cost $6. Write an expression to represent the cost of each purchase.

   a. 4 shirts _____

   b. 5 shirts and 9 posters
   _____

   c. n shirts and 6 posters
   _____

   d. 3 shirts and d posters
   _____

2. Trina has 6 more points than Alex in a video game. Write an algebraic expression for the number of points each person has earned. Use the same variable in each expression.

   a. Alex _____

   b. Trina _____

   c. Alex has 85 points. How many points does Trina have? Evaluate your expression to solve. Show your work.
   _____

3. Write an algebraic expression for each of the following. Use n as your variable. Then evaluate each expression for $n = 8$.

   a. the sum of 12 and a number
   _____

   b. the quotient of 24 divided by a number
   _____

4. Ribbon costs $3.00 per package. Which expression gives the total cost of p packages of ribbon? Circle the letter of the correct answer.

   A $p + 3$ B $3 - p$

   C $p \div 3$ D $3p$

5. Explain how to evaluate $4d$ for $d = 3$.

6. Write a word expression for each algebraic expression.

   $b + 8$ _____

   $b \div 2$ _____

Objective 11.2: Write expressions for situations that include an unknown quantity.

Lesson 11-3: Write Equations with Unknowns

Words to Know An **equation** is a statement that two quantities are equal. It is written with an equal symbol (=).

Activity 1

Write an equation to show the number relationship for this picture.

♡♡♡♡♡ + ? = ♡♡♡♡♡♡

5 + ? = 6

Practice 1

Write an equation to show the number relationship for this picture.

? − ☆☆☆☆☆ = ☆☆☆☆☆☆

____ − ____ = ____

Activity 2

Jake and his two brothers are at the skate park. Then some other friends joined them. Now there are 5 boys at the skate park. Write an equation that could be solved to find the number of friends who showed up.

Jake and his two brothers: 3
Some other friends: f
Total at the park: 5

$3 + f = 5$

Practice 2

There are 15 girls on the volleyball team. Some of them are out sick this week. There are 12 girls at the game. Write an equation that could be solved to find the number of girls out sick.

Girls on the volleyball team: _____

Girls out sick: _____

Total at the game: _____

_____

On Your Own

Jai brought some cans of food for the food drive. Karl brought 6 cans. Together, they brought 10 cans. Write an equation that could be solved to find the number of cans that Jai brought.

Write About It

Write a number story that could be shown by the equation $12 - \blacktriangle = 5$.

Objective 11.3: Write equations for word problems that include an unknown quantity.

Lesson 11-3: Write Equations with Unknowns
Understand It — B

Example 1
There are 12 students in the chess club. They each brought the same number of sandwiches to the club picnic. There were 36 sandwiches in all. Write an equation that could be solved to find the number of sandwiches each student brought.

$12 \times s = 36$

Practice 1
Each of the students in Mrs. Walter's class brought in 2 of their favorite pictures to make a class museum. There are 42 in the class museum. Write an equation that could be solved to find the number of students in Mrs. Walter's class.

Example 2
Scott baked 20 muffins for his aunts. He gave each of his 5 aunts the same number of muffins. Write an equation that could be solved to find the number of muffins Scott gave each of his aunts.

$20 \div m = 5$

Practice 2
Rita picked some flowers from her garden. She gave 2 flowers to each of her 4 friends. Write an equation that could be solved to find the number of flowers Rita picked.

On Your Own
Marta and her 3 friends each played in the same number of matches in the tennis tournament. They played in 40 matches all together. Write an equation that could be solved to find the number of matches Marta and her friends played in.

Write About It
Write a number story that could be described by this equation: $36 \div b = 4$.

Objective 11.3: Write equations for word problems that include an unknown quantity.

Lesson 11-3 Write Equations with Unknowns

1. Write a number sentence to show the number relationship for each picture.

   a.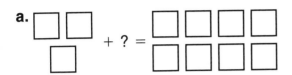

   _____

   b. ? − △△△△ = △△△

   _____

2. Kim had 3 clients. She added more clients to her list. Now she has 9 clients. Which equation can be used to find the number of clients Kim added to her list? Circle the letter of the correct answer.

   A $3 + c = 9$ **B** $3 - c = 9$

   C $c + 9 = 3$ **D** $3 + 9 = c$

3. John has a collection of 9 photographs. Julie gives him some more photographs for his birthday. Now he has 13 photographs. Write an equation that could be solved to find the number of photographs Julie gave John.

4. Kay had some students in her cooking class. Then 4 of them moved away. Now she has 10 students. Write an equation that could be solved to find the number of students that Kay had originally.

5. At Greta's party, an extra-large pizza was cut into 16 slices. Each person got 2 slices. Write an equation that could be solved to find the number of people at Greta's party.

6. Each member of the McGregor family brought 2 pies to the school bake sale. There were 8 pies all together. Write an equation that could be solved to find the number of people in the McGregor family.

7. Donatella volunteered to deliver 7 lunches to senior citizens. She delivered some. Now she has 3 lunches left. Which equation could be used to find how many lunches Donatella has already delivered? Circle the letter of the correct answer.

   A $3 + 7 = ?$ **B** $7 - ? = 3$

   C $? - 3 = 7$ **D** $? - 7 = 3$

8. Write a number story that could go with this equation: $y \times 3 = 18$.

Objective 11.3: Write equations for word problems that include an unknown quantity.

Volume 4   Level E

Lesson 11-4: Solve Equations with Unknowns

Model It — A

Words to Know An **equation** uses the equal sign to show that two expressions have the same value.

The **solution** to an equation is the value for the unknown that makes the equation a true statement.

Activity 1

$$y - 4 = 11$$

Think: what number less 4 equals 11? Since you must take 4 away from the value of y to get 11, y must be 4 more than 11.

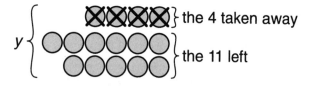

When 4 were taken away, there were 11 left, so there were 15 to begin with: $y = 15$.

Practice 1

Draw a picture to show $n - 3 = 14$.

What is the value of n?

_____

Activity 2

One box is covered. The total number of baseballs in both boxes is 10. How many baseballs are in the second box?

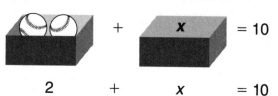

$$2 + x = 10$$

Think: what number added to 2 equals 10? The sum of 2 and 8 is 10, so there must be 8 baseballs in the covered box: $x = 8$.

Practice 2

Two boxes of golf balls are shown below. One is covered. The total number of golf balls in both boxes is 9.

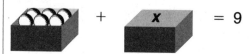

How many balls are in the open box? _____

How many golf balls must be in the closed box if the total is 9? _____

On Your Own

Draw a picture to help you find the solution to $z + 3 = 8$.

_____

Write About It

Use words to write the relationship shown in the following equation: $c - 7 = 12$. What is the value of c?

Objective 11.4: Write and solve simple equations for word problems that include an unknown quantity.

Lesson 11-4: Solve Equations with Unknowns

B — Understand It

Example 1

Karen had 15 photographs stored in her camera. After the class party, she had a total of 35 photographs stored. The equation below represents this story.

$$15 + ? = 35$$

To find the number of photographs Karen took at the party, use subtraction.

$$35 - 15 = 20$$

Karen took 20 photographs at the party.

Practice 1

Chato spent 20 minutes raking leaves in the front yard. Then he raked the back yard. After a total of 45 minutes, Chato was done with his chore. Write an addition equation to represent this story. Use a question mark to show the missing value.

_____

Find the number of minutes that Chato spent raking leaves in the back yard.

_____

Example 2

Before they opened, Darrell's Bakery baked some doughnuts. By 9 A.M., they had sold 62 doughnuts. There were 18 doughnuts left. The equation below represents this story.

$$d - 62 = 18$$

To find the number of doughnuts Darrell's Bakery baked, use addition.

$$62 + 18 = 80 \text{ doughnuts}$$

Practice 2

Ms. Wolski had some students in her beginner swimming class. 13 of them graduated to the advanced class. 24 remained in the beginner class. Write a subtraction equation to represent this story.

_____

Find the number of students who started out in the beginner swimming class.

_____

On Your Own

There are 320 paintings at the Modern Art Museum. All together, the Fine Art Museum and the Modern Art Museum have 610 paintings. Write an addition equation to represent this number story. Then find the number of paintings at the Fine Art Museum.

Write About It

Write a number story that could be represented by this equation. Then find the value for x.

$$100 - x = 8$$

Objective 11.4: Write and solve simple equations for word problems that include an unknown quantity.

Lesson 11-4 Solve Equations with Unknowns

1. Use mental math to find each missing value.

   a. ? − 30 = 50 _____

   b. ? + 5 = 25 _____

   c. 18 − ? = 6 _____

2. Use mental math or draw a picture to find each solution.

   a. 16 − e = 5 _____

   b. f + 8 = 13 _____

   c. 10 + g = 19 _____

   d. j − 11 = 6 _____

3. The park district plans to have 14 soccer teams this summer. They will also have some softball teams. All together, there will be 24 softball and soccer teams. Write an addition equation to represent this number story. Then find the number of softball teams.

   _____

4. Charlie had 35 music students. Some of them studied guitar. 26 studied piano. Write a subtraction equation to represent this number story. Then find the number of students who studied guitar.

5. Write a number story that could be represented by this equation. Then find the value for y.

   y + 30 = 45

6. Omar had $13. His aunt gave him some money for his birthday. Now Omar has $23. Write an addition number relationship using a ? to show how many dollars Omar's aunt gave him. Then solve.

Objective 11.4: Write and solve simple equations for word problems that include an unknown quantity.

Topic 11: Expressions and Equations

Topic Summary

Choose the correct answer. Explain how you decided.

1. Jill owns a dog-walking service. She had 8 dogs on her schedule today. She has 3 dogs left to walk. Which is **not** a way to represent this number relationship?

   A $8 - x = 3$

   B

   C

   D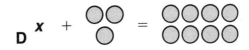

   _____
   _____
   _____

2. Val's Books is having a big sale. All paperbacks will sell for the same price. The price of a hardcover will be $7 more than the price of a paperback. If the price of a paperback is p dollars, which expression represents the cost of a hardcover?

   A p

   B $7p$

   C $p + 7$

   D $p - 7$

   _____
   _____
   _____

Objective: Review writing and solving expressions and equations in one variable.

Topic 11: Expressions and Equations

Mixed Review

1. Compare using < or >.

   a. 180 ◯ 108

   b. 86 ◯ 231

   c. 809 ◯ 908

   d. 131 ◯ 126

   Volume 1, Lesson 2-3

2. Find each product.

   a. 302 × 5 = _____

   b. 2,316 × 4 = _____

   c. 2,040 × 3 = _____

   d. 538 × 1 = _____

   Volume 4, Lesson 9-3

3. A hot dog stand sells 367 hot dogs a day. If the cost of a hot dog is $3, about how much does the hot dog stand make in one day? Circle the letter of the correct answer.

   A $900   B $1156.05

   C $1,200   D $1,600

   Volume 4, Lesson 9-2

4. Which shows six million, three hundred five thousand, nineteen in standard form? Circle the letter of the correct answer.

   A 635,019   B 6,350,019

   C 6,305,019   D 60,035,190

   Volume 1, Lesson 3-3

5. Find each sum. Show how you added.

   a. 6 + 2 + 7 = _____

   b. 3 + 9 + 4 = _____

   c. 1 + 5 + 7 = _____

   d. 8 + 6 + 9 = _____

   Volume 2, Lesson 4-2

6. Find the value of each expression for the given value of the variable.

   a. $a = 5$; $a + 2$ _____

   b. $b = 2$; $6b$ _____

   c. $c = 7$; $c - 4$ _____

   d. $e = 9$; $e \div 3$ _____

   Volume 4, Lesson 11-2

Objective: Maintain concepts and skills.

Words to Know/Glossary

C
compatible numbers — Numbers that are close to the actual numbers but easier to work with mentally.

D
dividend — The number you divide.

divisor — The number you divide by.

E
estimate — A number close to the actual answer.

M
mental math — Solving math problems in your head.

multiple — The product of a given number and another whole number.

Q
quotient — The answer to a division problem.

R
rounded — Rounded numbers have values close to the original amount that are often more convenient to use.

Word   **My Definition**   **My Notes**

Word **My Definition** **My Notes**

Index

A
addition
   using to solve equations, 41–43
algebraic expression, 35

D
division
   choosing a method for, 26–28
   of multidigit numbers by 1-digit numbers, 23–25
   of multiples of 10, 100, 1,000, 17–19
   problems, 29–31
   to solve equations, 44–46

E
equations
   definition of, 41, 44
   using addition and subtraction to solve, 41–43
   using multiplication and division to solve, 44–46
   writing, 38–40
estimating
   products, 5–7
   quotients, 20–22
expressions
   using unknowns in, 35–37

F
fewer than, 39

G
greater than, 39

I
inequalities, writing, 38–40

M
Mixed Review
   9: Use Multiplication to Compute, 23
   10: Use Division to Compute, 33
   11: Equations and Inequalities, 48
multiplication
   choosing a method for, 11–13
   of multidigit and 1-digit numbers, 8–10
   multiples of 10, 100, 1,000, 2–4
   problems, 29–31
   to solve equations, 44–46

P
products, estimating, 5–7

Q
quotients, estimating, 20–22

S
solution, 44
subtraction
   using to solve equations, 41–43

T
Topic Summary
   9: Use Multiplication to Compute, 22
   10: Use Division to Compute, 32
   11: Equations and Inequalities, 47

U
unknowns, 35–37

W
writing equations and inequalities, 38–40